THE
SOUTH EASTERN RAILWAY.

THE SOUTH EASTERN RAILWAY:

Its
Passenger Services, Rolling Stock,
Locomotives, Gradients,
and
Express Speeds.

BY

The Author of "British Railways." [Pattinson, J. P.]

WITH THREE PLATES.

LONDON,
PARIS & MELBOURNE:
CASSELL AND CO., LIMITED.
1895.

[ALL RIGHTS RESERVED.]

PREFACE.

THE present extended review of the older and more important of the lines to the Continent—the South Eastern Railway—practically brings up to date the article written three years since on this system in "British Railways." I was desirous in that volume of pointing out certain noticeably good features in the working of this much-abused line, and, as was to be expected, a few critics took exception to the favourable tone of my remarks. Further acquaintance with the South Eastern has confirmed me in the estimate already formed of the ability and resource displayed in the conduct of a passenger traffic of a more than usually embarrassing nature. The results of my observations, as set forth below, for convenience of comparison follow the arrangement adopted in "British Railways."

As the original article stood, it occupied only about half-a-dozen pages, and although several actual performances were briefly mentioned, only one was stated in detail. The review which follows is a much more exhaustive attempt to deal with the passenger services, rolling stock, and express speeds of the line. More than 150 instances of actual work are given, and at least 30 of them are in detail. These, it is hoped, will enable a more correct idea of the express work of the line to be formed than has, owing to the comparative paucity of published instances of actual performances, hitherto been possible.

I must here take the opportunity of acknowledging the kindness and attention of all the officials of the line with whom I came in contact during a two months' experience of the working of the system; and particularly must I thank Sir Myles Fenton for his assistance in many ways, without which it need hardly be said the fully detailed work now set forth would have been impossible.

CONTENTS.

(1) General Description of the Line	PAGE 1
(2) Travelling Facilities—	
(a) Services between Chief Towns	1
(b) Rolling Stock and General Accommodation	5
(3) Locomotive Work—	
(a) Speed	8
(b) Gradients	10
(c) Locomotives, &c.	13
(d) Actual Performances	15

INDEX OF PLATES.

I. Gradient Diagram: London to Hastings	*To face*	10
II., III. Speed Recorder Diagram: London to Hastings	,,	25

INDEX OF TABLES.

I. Travelling Facilities:

(a) Services between Chief Towns, August, 1892	3
(b) Services between Chief Towns, August, 1894	3

INDEX OF TABLES.

II. LOCOMOTIVE WORK:

PAGE

SPEED—

    (a) List of start-to-stop runs at or above 40 m.p.h. Summer of 1894 9

    (b) Summary of above 10

GRADIENTS—

    (a) Tunbridge Junction to Dover Town 11

    (b) Ashford to Canterbury 12

    (c) Dartford Line 12

LOCOMOTIVES, ETC.—

    (a) Dimensions 14

    (b) Weight of Passenger rolling stock 15

    (c) Dimensions of Boats 15

ACTUAL PERFORMANCES—

    (a) Tunbridge Junction to Dover. (Milepost timing.) . . 16

    (b) Continental Trains 18–22

    (c) Canterbury, Ramsgate, and Margate Trains . . 23–24

    (d) Hastings Trains 25–26

    (e) Chatham Trains 27–28

    (f) Uphill Running. New Cross to Halstead . . . 30

    (g) Best Performances 31

    (h) Summary of observations 32

EXPLANATORY NOTES.

(1) **General Description of the Line.**—In this section mention is made of the main routes and branches of the railway under review. The length of line is stated approximately, and includes only the lines actually owned by the Company, without reference to those leased or rented, or simply worked over.

(2) **Travelling Facilities.**—An examination of the opportunities and obligations of the system to run frequent and fast services of trains is here made. Tables showing the train services for the months of August, 1892 and 1894, between the principal points served by the Company are given. This month has been chosen so as to include the seaside trains running only in the summer. So far as can be judged at present, the figures for August, 1894, will not be materially altered for the coming summer of 1895. Punctuality and local train services are then discussed. Following this, under the sub-title of *Rolling Stock and General Accommodation*, the passenger carriages, safety appliances, stations, etc., of the Company are described.

(3) **Locomotive Work.**—This part of the subject is distinct from the foregoing, as the subject is no longer discussed from the traveller's point of view, but with a desire to consider in a general manner the work of the locomotive and the resistances to be overcome by it. Accordingly, under the separate sub-titles of (a) *Speed*, (b) *Gradients*, (c) *Locomotives*, (d) *Actual Performances*, we consider respectively (a) the demands made on the locomotive in the way of speed, (b) the contour of the line over which these speeds are to be maintained, (c) the machines actually doing the work indicated in the previous sections, and (d) the manner in which these machines actually perform this work, as illustrated by a very large number of

examples observed, personally, in actual daily practice. These examples furnish details as to the weight of the train, and—to avoid confusion and misapprehension—this is always exclusive of engine and tender; further, the types of passenger rolling stock being so numerous, the weight has been given in coaches of ten tons each. Thus, a train with six eight-wheeled coaches weighing twenty tons each would be described as a train of twelve coaches. By this means wearisome detailed description of the various types of carriages composing each train has been avoided.

The distances given in the section illustrating Actual Performances have been taken from authentic sources, and the figures representing the running of the various trains are, in every instance, the result of very careful personal observation.

THE
SOUTH EASTERN RAILWAY.

(1) General Description of the Line.

THE South Eastern Railway is over 380 miles in length, and possesses a very direct route to Dover and Folkestone (its ports for the Continent), through Sevenoaks and Ashford. At Dover the Company's trains connect with the boats of the Chatham, which cross to Calais, and with the fine service of the Belgian Government to Ostend; while at Folkestone the South Eastern own rather small but eminently seaworthy vessels, which cross to Boulogne, and thus provide a good alternative route to Paris. In addition to the main line to Dover, there are important branches from Tunbridge Junction to Tunbridge Wells and Hastings, and from Ashford to Canterbury, Ramsgate, and Margate. Besides these, there is a great number of minor lines, the most important of which are those from Reading to Tunbridge Junction, *viâ* Red Hill, and from London to Chatham and Maidstone, *viâ* Gravesend, at which point a line diverges to Port Victoria. Smaller branches connect Paddock Wood with Maidstone and with Hawkhurst, Ashford with Rye and Hastings, Canterbury with Whitstable, and Sandling with Hythe and Sandgate. The numerous ramifications of the system in the London district are very important.

(2) Travelling Facilities.

(*a*) SERVICES BETWEEN CHIEF TOWNS (*MONTH OF AUGUST*, 1894).—This line in many respects resembles its neighbour, the Chatham Company. Its efforts are largely concentrated on the

Continental traffic, which, with cross-Channel exclusiveness, is for the most part limited to **"first and second only,"** and for these classes there is a very fine service of expresses between London and Dover, as well as one express each way between London and Folkestone.

Trains like these leave little to be desired, the duration of the journeys between Cannon Street and Dover (75¼ miles) varying from about 100 to 115 minutes. Formerly, an express known as the "Club" train—avoiding Cannon Street—ran direct from Charing Cross to Dover. This and the corresponding up train were allowed 105 minutes for the journey, but both have now disappeared from the time-tables. Between Cannon Street and Folkestone (69¾ miles) the time allowed is 100 minutes down and 95 minutes up. On the down journey punctuality is, as a rule, well maintained; but the arrivals from the Continent can only, and we think fairly, be described as erratic—this, of course, being due to the late arrivals of the boats at Dover and to time lost in transferring mails and luggage there.

It will be noticed that Cannon Street is taken as the starting-point in stating the times of the above expresses as well as in the tables which follow. To Charing Cross the time is several minutes longer, and more on an equality with the London, Chatham and Dover from Victoria.

Looking now at the **third class services**, we find more to commend than is generally supposed to be the case. Very creditable expresses (one each way) run between London and Hastings, as shown below, but all the others fall below an inclusive rate of 35 miles an hour. Several trains to and from Canterbury are really fine expresses, and these, continued to Ramsgate and Margate, form a rather spirited competition with the shorter Chatham route, but in winter these towns are not well treated. To other points, the South Eastern service is poor, judged by the standards of the lines north of the Thames. The low inclusive rate of barely 35 miles an hour attained by the expresses to Chatham is, however, due to the number of stops made in so short a distance; while to Tunbridge Wells difficult gradients limit pace. The following tables show the state of South Eastern services for August, 1892, and August, 1894. Between these dates it will be seen that a few improvements have been effected. About the purely suburban services of London little can be said. Blackheath, Greenwich and Woolwich are well connected with the City—the rest, as regards frequency and speed of trains, are about the general English average.

AUGUST, 1892.

THIRD CLASS SERVICES.

Between London (Cannon Street) and	Miles.	No. of Trains at or over						Fastest.	Remarks.
		35 miles per hr.		40 miles per hr.		45 miles per hr.			
		Down.	Up.	Down.	Up.	Down.	Up.	Hr. Mn.	
Chatham . . .	31¾	0	2*	0	0	0	0	0 48*	* To London Bridge.
Tunbridge Wells .	33¼	1	0	0	0	0	0	0 54	Between London and the towns in this table, there are about four or five trains each way per day at from 30 to 35 miles per hour.
Hastings . . .	61½	1	1	0	0	0	0	1 37	
Canterbury . .	69¼	3	2	0	1	0	0	1 42	
Ramsgate . .	85	1	2	0	0	0	0	2 10	
Margate . . .	88¾	0	2	0	0	0	0	2 20	
Folkestone (Town) .	69¾	1	1	0	0	0	0	1 53	
Dover (Town) . .	75½	1	0	0	0	0	0	2 5	

AUGUST, 1894.

Between London (Cannon Street) and	Miles.	No. of Trains at or over						Fastest.	Remarks.
		35 miles per hr.		40 miles per hr.		45 miles per hr.			
		Down.	Up.	Down.	Up.	Down.	Up.	Hr. Mn	
Chatham . . .	31¾	0	2*	0	0	0	0	0 50	* To London Bridge.
Tunbridge Wells .	33¼	1	0	0	0	0	0	0 54	Other trains (both ways) at an inclusive speed of 30 to 35 miles per hour.
Hastings . . .	61½	1	1	0	0	0	0	1 37	
Canterbury . .	69¼	4	4	2	1	0	0	1 35	
Ramsgate. . .	85	4	4	2	0	0	0	2 4	
Margate . . .	88¾	4	4	0	0	0	0	2 17	
Folkestone (Town) .	69¾	2	1	1	0	0	0	1 40	Some other trains just fail to reach 35 miles per hour.
Dover (Town) . .	75½	2	1	1	1	0	0	1 51	

It is one thing to arrange train services on paper and another to carry them out in actual working; and if it be asked whether South Eastern arrivals approximate to the time-table issued by the Company, it must be admitted that at many points, more particularly at Charing Cross station, and on many occasions—such as times of exceptional traffic and on Saturday afternoons, when the station-working all over the line is slow—they do not. But the unpunctuality of the system is much less serious than is generally supposed; and, looking to the exceptional difficulties to be overcome, a good deal of it is excusable. Business people travelling habitually on a somewhat unpunctual line are naturally prone to over-estimate the precise amount of lateness. The writer, as a season-ticket holder on the Brighton line, began at one time to entertain strong ideas of what seemed to be the excessive unpunctuality of the arrivals at Victoria Station. In order to test the correctness of his surmises, he carefully noted down the times of nearly 150 consecutive arrivals of local trains by which he travelled. The average was about 4¾ minutes late each—a result by no means discreditable to the company. So, in repeated instances of apparently grave lack of punctuality on the South Eastern—with perhaps greater hindrances and congestion of traffic than fall to the share of the Brighton Company—it will probably be found that the arrivals do not average much more than 5 minutes behind booked time. We found—after noting 133 arrivals, including those of main-line, local and suburban trains—that the aggregate lateness amounted to 815 minutes, giving an average of just over six minutes per train.

From London Bridge to Charing Cross, *viâ* Cannon Street, is, without doubt, the most difficult piece of line in England over which to work a frequent service of trains. If this short stretch were eliminated, and the South Eastern had, like their Brighton neighbours, rested content with London Bridge as a terminus, the working of the trains would be carried on with much greater punctuality. The up trains reach London Bridge fairly to time on the average, and this notwithstanding the crowded state of the lines north of New Cross. Similarly, the down trains, once clear of London, rarely fail to maintain, and in some cases improve on, booked time, in spite of the discouraging effect a bad start exercises, as a rule, upon the staff in general and upon the engine-drivers in particular.

But on the Charing Cross side of London Bridge, and with such obstacles as Cannon Street and the loop outside, the railway management capable of preserving punctuality has not yet been found. If we count as two each train travelling *viâ* Cannon Street (for each arrives and departs on its way to or from Charing Cross), we have nearly 1,000 trains daily using this station—a number in excess of the figures for either Liverpool Street or Waterloo. Such congestion can, of course, only be relieved by widening the line, and in fact raises the question of finance as between public and shareholder.

It may with truth be said that the chief element in the unpopularity of the South Eastern is to be found in the enterprise with which it has carried its lines into the very heart of the City and West End. The traveller by the Brighton, alighting at that Company's London Bridge terminus, takes a cab or omnibus across the bridge to reach his place of business, or, if on foot, has frequently to do battle with the inevitable rain and mud of our English climate. Yet, curiously, he complains less than the South Eastern passenger who, at the cost of a few minutes' lateness, due to delays outside Cannon Street, reaches that station without change of vehicle, at the same or less expense, in about the same time, without possible weather discomforts and without bodily exertion—these better results being solely due to the fact that in the selection of a site for their City station the South Eastern have displayed more enterprise than their neighbours.

(*b*) ROLLING STOCK AND GENERAL ACCOMMODATION.—The passenger rolling stock of the South Eastern Railway is, like that of the other southern lines, too often scarcely up to modern requirements. Many new and commodious carriages, chiefly for the suburban traffic, have, however, been recently built, and in the course of a few years the coaching stock will doubtless be satisfactory in most respects. At present the first and second class carriages running on **the Continental and other express trains** are very comfortable, and (with the exception of the very latest firsts and seconds built by the Brighton Company) apparently superior to those employed by neighbouring lines. The upholstering and general fittings are often extremely tasteful. Lavatory accommodation is commonly provided, and to this Company belongs the credit of first affording such convenience to second-class passengers. For the less fortunate third-class passenger as much cannot be said. The carriage stock of this class—although in the case of recently-built vehicles tolerably roomy, and with fairly-proportioned window space and lofty roof—is still unprovided with cushions, except for the seat, and frequently divided into sets of two compartments separated by a low wooden partition. The older carriages have general dimensions and window space of the most meagre and primitive, but they do not appear to be so numerous in proportion to the total carriage stock as on the Chatham and Brighton lines, which, as a set-off, can show a few main-line thirds of a distinctly better type than any to be seen on the South Eastern.

The internal dimensions of main-line rolling stock vary so much in the third class that the average can hardly be stated, but in the first and second classes the following measurements may be taken as fairly representative:—

Height	7 ft. to 7 ft. 3 in.
Length	Do.
Width	About 7 ft. 6 in.

Turning now to the **local and suburban services**, we find that the first and second class carriages are, as a rule, very good, and differ but little from the main line standard. Some comparatively recently-built local, close-coupled trains, consisting of seven carriages, each mounted on two 8- or 12-wheeled bogie trucks, have improved the general average character of the rolling stock. The coaches of the third class, however, might with advantage be better upholstered. They are still considerably below the standard of the northern lines, but, except the oldest types, apparently superior to the stock of the neighbouring systems, the Brighton Company running on its suburban trains third class carriages open throughout, of very light build, and with unpleasantly rattling windows, and a great deal of the Chatham stock being quite as bad as the very worst South Eastern types.

As regards **special vehicles**, a few cars of Pullman type are run between London and Dover and Hastings. On the Continental trains several corridor firsts, splendidly fitted throughout (of which coach No. 2,125 may be taken as a favourable example), are in daily service, while the saloons running on the Dover and Folkestone boat expresses are probably the finest in the South of England.

The rolling stock is lighted either by the old oil lamp, or its modern substitute—gas. The old-style foot-warmers are used for heating purposes, and even these awkward apologies are more often than not absent in cold weather. In both these respects the Company merely follows ordinary English practice. The coaches lighted by gas would not be considered well illuminated by those accustomed to American methods. Some new trains, however, the first of which was turned out early in the present year, each consisting of thirteen coaches, are brilliantly lighted. Compressed oil gas, stored in cylinders under the framing, is used, and in the first and second class compartments the "Coligny Regenerative Lamp" is supplied. These carriages are also excellently and tastefully upholstered in the first and second classes, with thirds alike bright and smart in appearance, the light woodwork being relieved with dark panels. The trains are fitted with a new patent door, fastener (Kaye's patent), enabling the door to be opened from the inside, a distinct saving of time and trouble. This improvement is sure to be appreciated on the South Eastern, on which line, in common with the Midland, the door handles on the outside of the carriages are frequently stiff and hard to turn.

Station accommodation varies greatly. Sometimes good, sometimes bad, more frequently indifferent, the general standard cannot rank very high. In the London district, however, the South Eastern have been enterprising enough to establish the two most convenient and central stations—Cannon Street and Charing Cross—owned by any company, and the new London Bridge enlargement has been well carried out. Among the larger country stations which appear to fairly satisfy requirements may be mentioned Shorncliffe, Paddock Wood, Dorking, St. Leonards, and Hastings. On the other hand, Redhill, Reading, Margate, Ramsgate, and Chatham are decidedly poor, and much below the standard of the lines north of the Thames. Tunbridge also, and Ashford, appeared to me inadequate to traffic requirements. Most of the smaller stations on the main line are somewhat heavy-looking wooden structures, and often enough these, although giving fair accommodation, have been allowed to go without paint for a very considerable time, with manifest injury to the appearance of the wood-work. The colour used is a dull white, which might with advantage be relieved and heightened by the judicious employment of one or two other tints. On the other hand, at several small stations—and these, singularly enough, generally on branch lines—neglect of this sort was not conspicuous. Among these may be mentioned Gomshall, Chilworth, and Shalford, on the Reading line; Nutfield, near Redhill; Ticehurst Road and Battle, on the Hastings line; and one or two others. A pleasing feature is the decoration of stray stations here and there with flower-beds and plants. Smeeth, Wadhurst, and Sidcup are very noticeable examples. Over-bridges or subways are for the most part provided to ensure safety in crossing the line.

So far as could be seen, the permanent way is in excellent condition. Of the many millions of passengers on the South Eastern main line or branches, but few probably realise the amount of assiduous care bestowed on the structure and appearance of the road. The ballast is most regularly sloped, the hedge-rows, fencing and grass slopes carefully trimmed, tended, and maintained, and the telegraph poles kept in strict alignment. These may be trifles, but the largest railroad in the United States—the Pennsylvania—has in no small degree established its reputation by attention to them. As regards safety appliances, the line is well signalled, and the stock is furnished with the Automatic Vacuum brake. Its neighbours, on the contrary, have adopted the Westinghouse system. The electrical communication between passengers and guard has been employed for many years on the South Eastern, which may be congratulated on possessing a really reliable medium of communication. With this the cord system employed by the Northern Companies cannot for a moment compare.

(3) Locomotive Work.

(*a*) *Speed.*—Although there is very little booked speed on the South Eastern which exceeds 45 miles per hour, yet the locomotive work on this system is entitled to some considerable attention, and is probably quite as good as that of its neighbours, the London, Chatham and Dover, and Brighton Companies. Its chief characteristic is the hauling of tolerably fast and frequently very heavy trains over a main line much of which is on steep gradients. The Continental expresses have, as a rule, a booked speed of about 45 to maintain when once clear of London, and some slightly exceed this. Although in most cases limited to first and second classes only, they are generally heavy. The seaside trains run by the South Eastern have seldom booked speeds much above 40 per hour. The 10.5 a.m. and 1.15 p.m. down are (August, 1894), however, notable exceptions, having only the short allowance of 81 and 82 minutes respectively for the 65½ miles from New Cross to Canterbury.

The following table gives full particulars of every start-to-stop run in which a running average of 40 miles an hour is attained or exceeded. Runs of a shorter length than five miles are excluded, and the times are taken from the official working time-table of the Company and the distances from Airey's Clearing House maps. In cases where only one time is given for both arrival and departure, one minute has been assumed to be the length of stop, and has been allowed for. Runs at or over 45 miles per hour are in bolder type. Sunday trains are omitted. Details are given below.

AUGUST, 1894.

From	To	Booked Time of Departure.	Miles.	Minutes Allowed.
Ashford	Canterbury . . .	9 30 a.	14¼	20
,, ...	,, ...	9 43 a.	14¼	20
,, ...	,, ...	2 12 p.	14¼	20
,, ...	,, ...	4 48 p.	14¼	20
,, ...	,, ...	6 28 p.	14¼	20
,, ...	Cannon Street . .	8 58 a.	55	77
,, ...	Headcorn . . .	9 28 a.	10¾	16
,, ...	London Bridge . .	10 40 a.	54¼	79
,, ...	Paddock Wood . .	5 6 p.	21¼	29
,, ...	,, ...	5 20 p.	21¼	29
,, ...	Shorncliffe . . .	11 5 p.	13¼	19
,, ...	,, ...	11 50 p.	13¼	19
,, ...	Smeeth . . .	9 32 a.	9¼	13
,, ...	Tunbridge Junction .	12 55 a.	26½	38
,, ...	,, ...	2 40 p.	26½	38
Battle	Etchingham . . .	8 21 a.	8	12
,, ...	,, ...	12 1 p.	8	12
Cannon Street . .	Ashford	11 10 a.	55	75
Cannon Street . .	**Dover Town** . .	**8 10 a.**	**75½**	**98**
Cannon Street . .	Dover Town . .	8 25 p.	75½	102
Cannon Street . .	**Dover Town** . .	**11 7 a.**	**75½**	**96**
,, ...	**Folkestone** . . .	**10 8 a.**	**69¾**	**91**
Cannon Street . .	Sandling . . .	5 44 p.	64¼	87
,, ...	West St. Leonards .	3 50 p.	59½	88
Canterbury . . .	Ashford	8 35 a.	14¼	20
,, ...	,, ...	10 16 a.	14¼	21
Canterbury . . .	**Minster** . . .	**2 34 p.**	**11¼**	**15**
,, ...	,, ...	**5 10 p.**	**11¼**	**15**
,, ...	,, ...	**6 50 p.**	**11¼**	**15**
Canterbury . . .	New Cross . . .	7 50 p.	65½	89
,, ...	,, ...	2 40 p.	65½	89
Dorking . . .	Reigate . . .	10 15 a.	6¼	9
,, ...	,, ...	2 43 p.	6¼	9
Dover Town . . .	Cannon Street . .	4 0 a.	75½	104
,, ...	,, ...	2 45 p.	75½	102
,, ...	Staplehurst . . .	1 50 a.	34¾	49
Folkestone . . .	**Cannon Street** . .	**4 10 p.**	**69¾**	**93**
Folkestone . . .	**Dover** . . .	**1 8 p.**	**5¾**	**7**
Gravesend . . .	London Bridge . .	12 10 p.	22	33
,, ...	,, ...	5 47 p.	22	33
,, ...	,, ...	10 18 p.	22	33
,, ...	Strood	10 36 a.	7¼	10
Grove Ferry . . .	Minster	7 53 p.	5	7
Grove Ferry . . .	**Minster** . . .	**11 9 a.**	**5**	**6**
Headcorn . . .	**Ashford** . . .	**6 10 p.**	**10¾**	**14**
London Bridge . .	Ashford	3 28 p.	54¼	77
,, ...	,, ...	3 44 p.	54¼	77
Minster . . .	Grove Ferry . . .	7 58 a.	5	7
New Cross . . .	**Canterbury** . . .	**10 27 a.**	**65½**	**81**
,, ...	,, ...	**1 38 p.**	**65½**	**82**
Paddock Wood . .	**Staplehurst** . .	**5 52 p.**	**7**	**9**

(*Continued on next page.*)

From	To	Booked Time of Departure	Miles.	Minutes Allowed.
Red Hill	London Bridge	9 0 a.	20¾	29
"	"	9 55 a.	20¾	30
"	"	10 31 a.	20¾	30
Sandling	Ashford	7 27 p.	9¼	13
"	"	9 12 a.	9¼	13
"	Cannon Street	9 53 a.	64¼	89
Sandling	**Cannon Street**	**5 45 p.**	**64¼**	**84**
Sevenoaks	Hildenborough	10 41 p.	5	7
Shorncliffe	Ashford	10 36 p.	13¼	19
Staplehurst	"	11 22 p.	14¼	20
"	Tunbridge Junction	2 41 a.	12¼	17
Tunbridge Junction	**Ashford**	**9 7 a.**	**26½**	**34**
Tunbridge Junction	Ashford	1 28 p.	26½	39
"	"	3 24 p.	26½	37
Tunbridge Junction	**Ashford**	**5 33 p.**	**26½**	**35**
Tunbridge Junction	Ashford	8 8 p.	26½	38
"	Staplehurst	11 0 p.	12¼	18
Wadhurst	Etchingham	10 21 p.	8¼	12
"	West St. Leonards	6 8 p.	21¼	32
Wokingham	Reading	6 43 p.	6¾	10
"	"	7 47 p.	6¾	10

The table summarises as under :—

No. of Runs.	Description.	Mileage.	Minutes.	Speed m.p.h.
56	**40–45 m.p.h.**	**1467**	**2065**	**42·58**
16	**45–50 m.p.h.**	**601**	**775**	**46·53**
72	**All over 40 m.p.h.**	**2068**	**2840**	**43·69**

(*b*) *GRADIENTS.*—The South Eastern main line is—like the Chatham and Dover—a heavy course. The gradients, however, occur differently, and are met with all at once, instead of being short yet incessant as on the other route. The accompanying diagram of the gradients from London to Hastings, on a scale of 300 ft. vertical to the inch, shows the profile of a very severe section. The main line trains to Dover have, of course, to encounter the same heavy grades north of Tunbridge Junction, from which point eastwards the running is much easier, the grades in detail being as under, those of exceptional length and of any difficulty being shown in bolder type :—

Plate I.

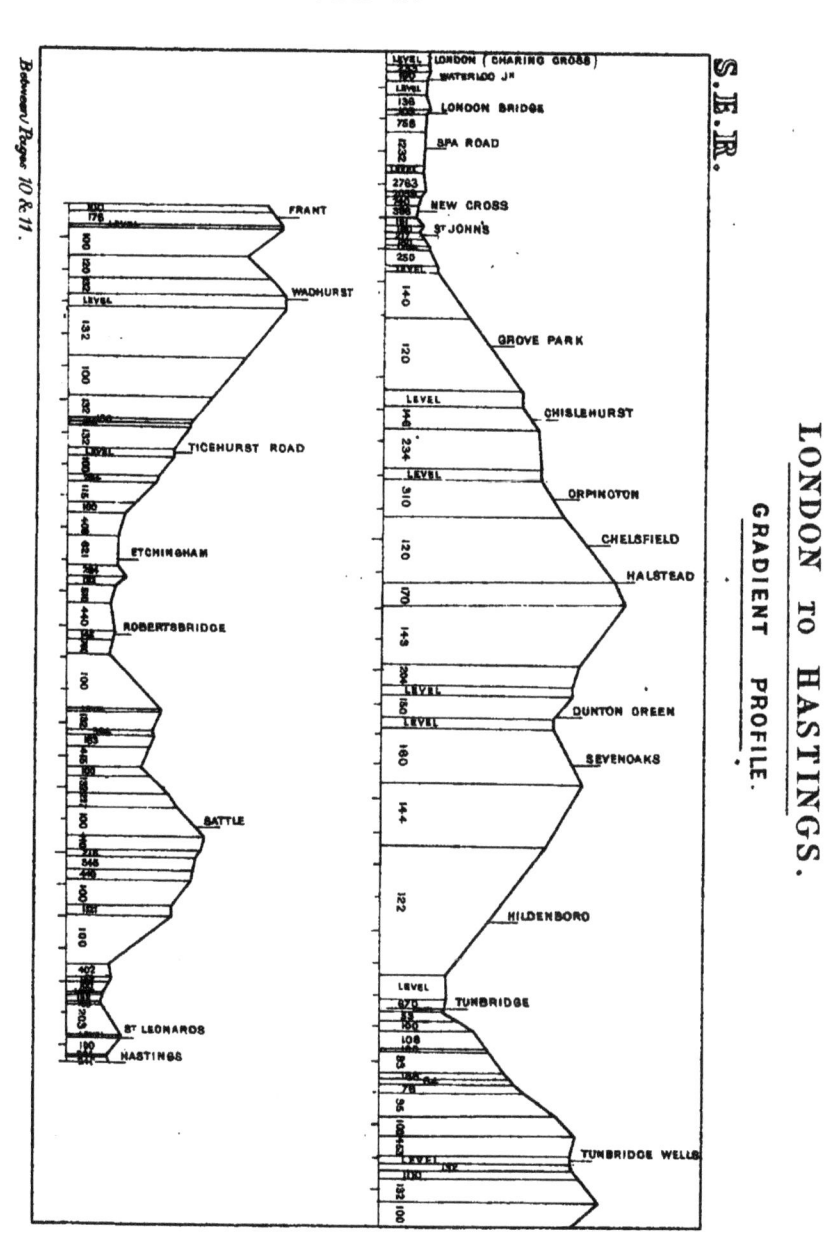

LOCOMOTIVE WORK.

GRADIENTS FROM TUNBRIDGE JUNCTION TO DOVER TOWN.

Rate of Gradient.	Length of Gradient.	Up or Down.	Remarks.	Rate of Gradient.	Length of Gradient.	Up or Down.	Remarks.
1 in.	mls. yds.			1 in.	mls. yds.		
670	0 550	Down		533	0 1320	Up	
258	1 550	Up		2026	0 374	Down	
446	0 330	Up		277	1 286	Up	
285	0 1562	Down		344	0 264	Down	
260	1 528	Down		5979	0 440	Up	
1164	0 1628	Down		338	0 770	Down	
288	0 682	Down		1699	0 396	Up	
690	0 660	Down	Paddock Wood at beginning of this grade.	359	0 440	Down	
				9149	0 1210	Down	
350	0 660	Up		273	0 770	Down	
325	0 440	Down		2260	0 1540	Up	Ashford at beginning of this grade.
578	1 0	Up					
Level	0 660	Level		259	1 440	Up	
324	0 440	Down		1671	0 1034	Up	
599	0 440	Down		342	1 616	Up	
402	0 770	Up		1507	0 374	Down	
490	0 946	Up		368	0 506	Down	Smeeth near middle of this grade.
252	0 1144	Up	Marden near middle of this grade.	655	0 330	Down	
388	0 968	Up		266	2 0	Up	
259	0 1496	Down		286	1 1430	Up	Westenhanger near end of this grade.
2435	0 330	Down					
537	0 1100	Up	Staplehurst near middle of this grade.	266	0 1320	Down	
395	1 44	Down		8408	0 440	Down	
				266	2 440	Down	
2102	2 220	Up		235	0 770	Down	
620	0 880	Up	Headcorn near beginning of this grade.	263	1 374	Down	Shorncliffe near middle of this grade.
2086	0 770	Down		381	0 462	Down	
451	0 1540	Up		259	0 814	Down	
244	0 1100	Up		2760	0 262	Down	
343	0 1276	Up		375	0 616	Down	
1628	0 660	Down		264	4 220	Down	Folkestone at beginning of this grade.
266	1 0	Up					
651	0 374	Down		553	0 220	Down	
329	0 1100	Down		238	0 1430	Down	
410	0 770	Down	Pluckley near beginning of this grade.	182	0 220	Down	
287	1 0	Up		Level	0 418	Level	

The other sections over which fast or express trains run are those from Ashford to Ramsgate and the line from London to Chatham, *viâ* Gravesend and Strood. Details of the former as far as Canterbury, and of the latter from the junction with the main line as far as Dartford, are given below. From Canterbury to Ramsgate the grades, with the exception of 2¼ miles of 113 and 100 rising before St. Lawrence, are easy, mostly undulating at 1 in 264 and 330; while the same may be said of the profile between Dartford and

Strood, except before Greenhithe, where we find ¼ mile of 1 in 211 rising, 1 mile of 1 in 182 rising, ¾ mile of 1 in 129 falling, and 1½ miles of 1 in 210 rising.

ASHFORD TO CANTERBURY.

Rate of Gradient.	Length of Gradient.	Up or Down.	Remarks.	Rate of Gradient.	Length of Gradient.	Up or Down.	Remarks.
1 in.	mls. yds.			1 in.	mls. yds.		
400	0 1276	Up	Joins main line at beginning of this grade.	220	0 968	Down	
264	1 880	Down		660	1 660	Down	Chilham at beginning of this grade.
Level	0 1122	Level		132	0 440	Up	
264	0 462	Up		132	0 660	Down	
264	0 1166	Down		400	1 220	Down	Chartham at beginning of this grade.
660	0 1254	Down	Wye at beginning of this grade.	550	0 1144	Down	
300	0 1320	Up		1320	0 1056	Down	
200	0 1100	Down		220	0 1012	Up	
132	0 1320	Up		132	0 462	Down	
200	0 880	Down		Level	0 176	Level	
176	0 1232	Down		600	0 330	Up	Canterbury near middle of this grade.

DARTFORD LINE.

Rate of Gradient.	Length of Gradient.	Up or Down.	Remarks.	Rate of Gradient.	Length of Gradient.	Up or Down.	Remarks.
1 in.	mls. yds.			1 in.	mls. yds.		
160	0 110	Up	Leaves main line at beginning of this grade.	100	0 1430	Down	
200	0 110	Down		Level	0 440	Level	Bexley in middle of this grade.
Level	0 396	Level		132	0 1408	Down	
180	0 1364	Up	Lee near middle of this grade.	Level	0 726	Level	
120	0 572	Up		198	0 594	Down	Crayford ¼ way along this grade.
250	1 88	Up	Eltham near end of this grade.	330	0 1320	Up	
			Pope St. ¼ way and Sidcup at end of this grade.	Level	0 660	Level	Joins line to Strood at end of this grade.
1800	2 440	Down		4257	0 1386	Up	Joins line from Crayford ½ way.
165	0 1320	Down		Level	0 638	Level	Dartford near end of this.
Level	0 440	Level					

Other parts of the South Eastern system, over which no express trains run, are by no means free from very heavy gradients. A few notes on these are here appended:—

Sandling to Sandgate is nearly all 1 in 56, 54, and 59 down.

There are 1,100 yards of 1 in 30 down to Folkestone Harbour.

Strood to Maidstone is almost level, but has 1¼ miles of 1 in 178 rising after Aylesford.

There is 1 mile of 1 in 66 rising between Cliffe and Sharnal Street.

There are 3 miles of 1 in 71, 72, 64, and 70 from before Walmer to Martin Mill; and 2¼ of 1 in 70 down to Dover on other side.

The Whitstable line, according to official gradient diagrams, rises 2 miles of 1 in 76, 41, 47, 56, 49 from Canterbury through tunnel, then 1⅛ miles of 1 in 843 up, then 836 yards fall at 1 in 31, and 594 yards at 1 in 28, then 1 mile 330 yards level, ⅓ mile fall at 1 in 57 and 50, and ½ mile nearly level. The trains are very light on this branch, and are worked by a tank engine.

On the Ashford to Hastings line Ham Street is in centre of 2 miles of 1 in 100 falling (to Hastings); and 3 miles after Winchelsea commences a 4 mile stretch up at 1 in 100, followed by 1 mile down at 1 in 132, and 1 at 1 in 60.

There are 8 miles of 1 in 264 up, ending at the northern mouth of Merstham Tunnel, beginning just north of East Croydon

Easy down grades prevail between Red Hill and Tunbridge. Reigate to Boxhill is down, chiefly 1 in 142, 116, and 125; then follows a stretch of 4 miles up at 1 in 96 chiefly, and afterwards 7 miles mostly 1 in 100 down, to Shalford.

There is an ascent of about 2 miles of 1 in 100 west of Guildford.

The single line from Purley to Caterham is 1 in 160, 118, 190 up after Kenley.

(*c*) *LOCOMOTIVES, etc.*—The South Eastern locomotives are well fitted for the hard main-line work described in the sections on Speed and Gradients. The works of the Company are at Ashford, where the locomotive stock is under the care of Mr. James Stirling and the carriages and wagons are looked after by Mr. W. Wainwright. The dimensions of the principal classes employed in fast train working are stated in the following table :—

Dimensions.		7 ft. coupled Bogie Express.	6 ft. coupled Bogie Express.	Four-coupled type.	Four-coupled type.	Latest type of Tank Engine.
Cylinders, Diameter . . . in.		19	18	17	16	18
,, Stroke . . . ,,		26	26	24	24	26
Heating Surface, Total . . sq. ft.		984·26	922·5	982	956	948·5
Fire-grate Area . . . ,,		16·78	15·5	15·8	14.8	15
Wheels, Leading, Diameter . ft. in.		3 9*	3 8*	4 0	4 6	5 6
,, Driving, ,, . . ,,		7 0	6 0	6 6	6 0	5 6
,, Trailing, ,, . . ,,		7 0	6 0	6 6	6 0	3 9*
,, Tender, ,, . . ,,		4 0	3 8	3 6	3 8	—
Weight on Leading Wheels . tons, cwts.		13 12*	12 12*	9 10	9 15	13 17
,, ,, Driving ,, . ,, ,,		15 18	14 2	14 10	12 15	16 0
,, ,, Trailing ,, . ,, ,,		13 0	11 5	10 5	9 15	18 16*
,, Total . . . ,, ,,		42 10	37 19	34 5	32 5	48 13
Capacity of Tank . . . gals.		2,650	2,000	2,500	1,600	1,050
Fuel Space, Coal . . . tons.		4	3	3	2	—
Weight of Tender . . tons, cwts.		30 10	25 14	30 15	20 10	—
						* Bogie.

The large 19-inch express engines were first built in 1884. One of them was shown in the Paris Exhibition of 1889, and attracted much attention. A full description of it appeared in the *Revue Générale des Chemins de Fer* for July, 1890. The general appearance is handsome, these engines being painted black with red stripes, though one or two have recently been painted a dark green. Three of the class originally had 18-inch cylinders. The design, in many particulars, calls to mind several of the earlier types of the Glasgow and South Western Company. Among such similarities may be mentioned the rounded shape of the cab, and the absence of a dome. They perform very good work in actual practice, and have earned a reputation for hauling heavy loads up steep grades without taking pilot assistance. The Continental, Hastings, and Ramsgate and Margate expresses are worked by these engines, while the Chatham expresses are worked by the old 6-feet coupled type, the new 6-feet coupled bogies, and the class of tanks described in the table above. Examples of the performances of these three smaller types will be found in detail in the runs illustrating the work of the trains between London and Chatham, under the heading of "Chatham Trains."

The official weights of several types of passenger rolling stock generally used in fast-train service may with advantage be given here :—

	Tons.	cwts.
(a) Cars running to Dover and Hastings	23	6
(b) Eight-wheeled Bogie Carriages	19	12
(c) Six-wheeled Saloon	14	8
(d) Six-wheeled Lavatory Carriage	13	0
(e) Six-wheeled Thirds, with six compartments	13	0
(f) Six wheeled Guard's Van	11	10
(g) Four-wheeled Guard's Van	9	15

The boats used in the Folkestone-Boulogne service are, with a few dimensions, as under :—

	Length.	Breadth.	Gross Tonnage.
Albert Victor	250	29	814
Louise Dagmar	250	29	816
Mary Beatrice	255	29	817
Duchess of York	270	30	Building.

Besides these there are four cargo boats employed on this route, and two ferry steamers, plying between Port Victoria and Sheerness, are also owned by the Company.

(d) ACTUAL PERFORMANCES.—So far as the writer is aware, scarcely any observations of the actual daily work of South Eastern locomotives have, up to the present time, been published, either in the few books dealing with this subject or in the considerable mass of correspondence which has from time to time appeared in the columns of the *Engineer*, the *English Mechanic*, the *Railway World*, and other papers which have opened their columns to the discussion of train speeds. By the aid of facilities, however, kindly placed at the writer's disposal by the management, over a hundred and fifty records of work from actual practice have been gathered together, thus illustrating what South Eastern locomotives do more fully than has ever previously been attempted. It is proposed to classify these into four sets as under :—

(a) CONTINENTAL TRAINS.
(b) CANTERBURY, RAMSGATE AND MARGATE TRAINS.
(c) HASTINGS TRAINS.
(d) CHATHAM TRAINS.

In section (a) 12 of the best performances are illustrated in detail, followed by 8 in semi-detail, and 79 further instances are given more briefly. In section (*b*) 19 performances are given, 4 of them in detail; in (*c*) 23 performances, 2 of them in detail, and 2 more in the form of a speed recorder diagram; and in (*d*) 13 performances, 3 of them in detail. Following these illustrations of actual practice, the subject of speeds on down-grades is considered, and then some considerable space is devoted to the uphill running of the South Eastern locomotives, concluding with a statement of the record performances between each pair of stations on the main line to Dover, and a summary of experiences.

With the exception of the runs illustrating the work of the trains to Chatham, the whole of the records which follow are performances of Mr. Stirling's 19-inch class of locomotive. The number of the engine has, wherever possible, been given. The runs, almost without exception, were made in the months of July, August, and December, 1894, and the booked times of departure, which in many cases are not quite the same as those stated in Bradshaw and the public time tables of the company, are taken from the official Working Time Tables for those months.

(*a*) CONTINENTAL TRAINS.

On the main line the work done is sufficiently shown by the numerous details given in the tables which follow. Particularly noticeable are the really excellent runs from Tunbridge Junction to Ashford with heavy trains, grades being, on balance, slightly against the train. Before coming to the tables it may be well to emphasise this point by illustrating a run on the 11 a.m. down. The times from post 41 (which is about half a-mile east of Tunbridge Junction) to post 86¾ are here presented:—

ENGINE 130.—TUNBRIDGE JUNCTION TO ASHFORD, 21½ COACHES; THENCE TO DOVER, 17 COACHES.

Mile-posts.	Time between Mile-posts.	Mile-posts.	Time between Mile-posts.	Mile-posts.	Time between Mile-posts.	Mile-posts.	Time between Mile-posts.
	Secs.		Secs.		Secs.		Secs.
41		52	70	64	74	78	190
42	87	53	68	65	76	79	60
43	84	54	65	66	69	80	59
44	73	55	67	68	134	82	129
45	67	57	136	69	70	83	69
46	67	58	72	70	70	86¾	215
47	69	59	76	71	73		
48	70	60	76	72	70		
49	69	61	70	73	73		
50	74	62	70	74	75		
51	75	63	73	75	77		

Even with the fine running indicated above, the booked time of this particular express was exceeded by over four minutes between London and Dover, and it may here be worth while to point out that of the 20 performances of Continental expresses given below in detail, in no fewer than ten cases time was lost even after allowing for signal checks. This is, without doubt, unusual, and, at first sight, it might appear that the locomotives were not up to their work. Such an assumption is, however, at once negatived when we come to examine the record of the fast Ramsgate and Margate expresses (*see* Runs Nos. 100 to 103), which, contrary to expectation, showed the best work we noticed. Here, with trains of 12½ and 17 coaches, the time from passing St. John's to passing Pluckley was better by more than 3 minutes than was observed with any of the Continental trains. A further proof, if required, is afforded by the work detailed in down run No. 4 and up runs Nos. 7 and 11.

The real reasons for the loss of time in running may be summed up as follows:—In cases 3, 5, 6, and 14 (down) the loads were extremely heavy (much heavier than most English locomotives would be expected to haul, unassisted, on up-grades of 1 in 120), and this, coupled with "greasy" weather caused booked time to be slightly exceeded. This explanation also holds good as regards run 17 (up), which had, in addition, to contend with a heavy gale from the west. A further explanation is to be found in the fact that South Eastern locomotives do not make up time by hurrying down the banks, a speed of 60 miles an hour being rarely exceeded. Finally, one can hardly judge the running of the up Continental expresses by the usual standard. Starting almost always very late from Folkestone and Dover, there is not that inducement to push the engine to its utmost that would be found were the train running to time. With a train already perhaps half or three-quarters of an hour late, a minute or so more dropped in running is not much noticed.

It may be well here to point out that, in the tables which follow, the figures indicating the time taken on each journey are exact to a second, whereas those giving the time lost by signal delays, etc., are, of necessity, although most carefully estimated, probably only correct within limits of a few seconds, and have, consequently, been stated to the nearest ¼ minute. The net time, therefore, can only, in those cases where time for signal delays has been deducted, be considered as accurate within the limit specified.

18   THE SOUTH EASTERN RAILWAY.

	Run No. 1.	No. 2.	No. 3.	No. 4.	No. 5.	No. 6.
Train	11 7 a.m. (Sundays).	11 7 a.m. (Sundays).	11 7 a.m.	10 8 a.m. (Sundays).	8 10 a.m. (Sundays).	8 10 a.m. (Sundays).
From	Cannon Street.	Cannon Street.	Cannon Street.	Cannon Street.	Cannon Street.	Cannon Street.
To	Sandling.	Sandling.	Dover Town.	Folkestone.	Folkestone.	Folkestone.
Due to Arrive	12 29 p.m.	12 29 p.m.	12 43 p.m.	11 39 a.m.	9 38 a.m.	9 38 a.m.
Engine	210	74	156	2	241	138
Load (Coaches)	13¼	17	20	15	18½	19
	H. M. S.	H. M. S.	H. M. S.	H. M. S.	H. M. S.	H. M. S.
Actual Departure	11 8 50	11 13 56	11 8 18	10 7 50	8 11 56	8 12 18

	Distance.	Time from Start.	Time from Start.	Time from Start.	Time from Start.	Time from Start.	Time from Start.
	M. Ch.	Min. Sec.	Min. Sec.	Min. Sec.	Min. Sec.	Min. Sec.	Min. Sec.
Cannon Street							
London Bridge	0 55	1 40	1 48	1 40	1 48	1 35	1 37
Spa Road	1 50	3 23	3 37	3 23	3 30	3 21	3 13
New Cross	3 59	6 24	6 52	6 20	6 23	6 15	6 2
St. John's	4 32	7 23	7 55	7 18	7 20	7 9	6 57
Grove Park	7 62	12 54	13 38	12 59	12 46	12 38	12 10
Chislehurst	10 10	17 16	18 3	18 5	17 35	17 27	16 47
Orpington	12 52	21 23	22 16	22 41	21 59	21 51	21 10
Chelsfield	14 6	23 35	24 29	25 4	24 18	24 3	23 27
Halstead	15 29	25 56	26 55	27 44	26 43	26 26	25 57
Dunton Green	19 27	30 48	32 2	33 6	31 41	31 34	31 32
Sevenoaks	20 70	32 31	33 44	34 55	33 28	33 23	33 25
Hildenborough	25 62	38 0	39 15	40 46	39 16	39 12	39 47
Tunbridge Junc.	28 29	40 29	41 57	43 28	41 54	41 47	44 34
Paddock Wood	33 52	46 29	47 46	49 33	47 51	47 54	53 46
Marden	38 15	51 28	52 42	54 23	52 39	52 59	58 55
Staplehurst	40 56	54 16	55 31	57 13	55 27	55 59	61 50
Headcorn	44 3	57 47	59 1	60 47	58 53	59 44	65 30
Pluckley	49 25	63 45	65 2	67 4	64 52	66 23	71 43
Ashford	54 71	70 1	71 22	73 43	71 17	73 32	78 20
Smeeth	59 16	74 45	76 22	78 30	76 7	78 44	83 22
Westenhanger	62 77	79 18	81 4	83 0	80 45	83 43	88 10
Sandling	64 18	81 5	82 53	84 33	82 21	85 27	89 49
Shorncliffe	68 14			88 27	86 27	89 35	93 55
Radnor Park	68 68			89 9	87 14	90 24	94 42
Folkestone	69 59			90 2	88 24	91 32	95 56
Dover	75 32			96 14			

	H. M. S.	H. M. S.	H. M. S.	H. M. S.	H. M. S.	H. M. S.
Actual Arrival	12 29 55	12 36 49	12 44 32	11 36 22	9 43 30	9 48 14
Time Taken	81 5	82 53	96 14	88 24	91 32	95 56
Signal Delays		0 30				6 0
Net Time	81 5	82 23	96 14	88 24	91 32	89 55
Time Allowed	82 0	82 0	96 0	91 0	88 0	88 0
Gain or Loss	Gain 55 sec.	Loss 23 sec.	Loss 14 sec.	Gain 2 min. 36 sec.	Loss 3 min. 32 sec.	Loss 1 min. 56 sec.
Particulars of Signal Delays, etc.		½ min. delay through fog in London.	Load was 15½ only from Ashford (Coaches slipped).			Checked at south end of Sevenoaks Tunnel and again (with a short stop) outside Tunbridge Junction.

		No. 7.	No. 8.	No. 9.	No. 10.	No. 11.	No. 12.
Train		3 4 p.m. (Sundays).	3 4 p.m. (Sundays).	5 45 p.m.	4 10 p.m. (Sundays).	4 10 p.m. (Sundays).	4 10 p.m. (Sundays).
From		Shorncliffe.	Shorncliffe.	Sandling.	Folkestone.	Folkestone.	Folkestone.
To		London Bridge.	London Bridge.	Cannon Street.	Cannon Street.	Cannon Street.	Cannon Street.
Due to Arrive		4 31 p.m.	4 31 p.m.	7 9 p.m.	5 42 p.m.	5 42 p.m.	5 42 p.m.
Engine		240	140	139	19 in. type.	140	224
Load (Coaches)		13	13	15	13¼	16¼	12¼
		H. M. S.	H. M. S.	H. M. S.	H. M. S.	H. M. S.	H. M. S.
Actual departure		3 39 4	3 20 19	6 3 49	4 26 3	4 12 1	4 35 0

	Distance.	Time from Start.	Time from Start.	Time from Start.	Time from Start.	Time from Start.	Time from Start.
	M. Ch.	Min. Sec.	Min. Sec.	Min. Sec.	Min. Sec.	Min. Sec.	Min. Sec.
Folkestone							
Radnor Park	0 71				2 30	2 0	2 25
Shorncliffe	1 45				3 51	3 16	3 39
Sandling	5 41	6 49	6 9		10 9	9 3	9 45
Westenhanger	6 62	8 31	7 51	2 43	11 57	10 43	11 30
Smeeth	10 43	13 21	12 1	7 37	16 23	15 9	16 4
Ashford	14 68	17 27	16 39	12 57	21 6	19 59	20 50
Pluckley	20 34	23 46	23 33	19 56	27 45	26 43	27 28
Headcorn	25 56	29 39	29 25	25 58	33 42	32 34	33 18
Staplehurst	29 3	33 16	33 22	29 44	37 31	36 12	37 0
Marden	31 44	36 18	36 49	32 46	40 45	39 14	39 59
Paddock Wood	36 7	41 26	42 16	37 53	46 43	44 31	45 9
Tunbridge Junc.	41 30	48 4	48 40	44 49	53 22	51 25	51 40
Hildenborough	43 77	51 41	52 27	48 44	57 28	55 26	55 22
Sevenoaks	48 69	60 9	61 1	57 10	66 12	65 3	64 5
Dunton Green	50 32	62 54	62 47	59 2	68 2	66 53	65 51
Halstead	54 30	66 56	68 44	64 27	73 44	72 37	71 25
Chelsfield	55 53	68 16	70 22	65 57	75 14	74 6	
Orpington	57 7	69 41	71 51	67 35	76 47	75 36	74 31
Chislehurst	59 49	72 16	74 34	70 22	79 29	78 2	77 15
Grove Park	61 77	74 34	76 59	72 57	82 0	80 16	79 48
St. John's	65 27	78 13	80 35	76 37	85 56	83 42	83 32
New Cross	66 0	79 6	81 25	77 28	86 48	84 29	84 27
Spa Road	68 9			80 21	89 29		
London Bridge	69 4	83 23	85 36	81 52	91 5	88 41	89 4
Cannon Street	69 59			84 19	92 55	90 29	90 49

		H. M. S.	H. M. S.	H. M. S.	H. M. S.	H. M. S.	H. M. S.
Actual Arrival		5 2 27	4 45 55	7 28 8	5 58 58	5 42 30	6 5 49
Time Taken		83 23	85 36	84 19	92 55	90 29	90 49
Signal Delays				0 45	0 15		
Net Time		83 23	85 36	83 34	92 40	90 29	90 49
Time Allowed		87 0	87 0	84 0	92 0	92 0	92 0
Gain or Loss		Gain 3 min. 37 sec.	Gain 1 min. 24 sec.	Gain 26 sec.	Loss 40 sec.	Gain 1 min 31 sec.	Gain 1 min. 11 sec.
Particulars of Signal Delays, etc.				Outside Cannon St. Station.	½ min. Signal Check outside Cannon St. Gale from west all the way.		

20   THE SOUTH EASTERN RAILWAY.

No. of Run.	DOWN. From	To	Train.	Actual time left.	TIME TAKEN TO PASS						Actual time arrived.	Time taken.	Time allowed.	Lost by signal checks.	Net Time.	Distance.	Engine.	Load.	Remarks.
					New Cross.	Chislehurst.	Tunbridge Junction.	Halstead.	Tunbridge Junction.	Ashford.						M.Ch.		
					H. M. S.	M. S.	M. S.	M. S.	M. S.	M. S.	H. M. S.	M. S.	M.	M. S.	M. S.	M.Ch.		
13	Cannon St.	Shorncliffe	8 25 p.m.	8 26 10	6 40 18	28 27	59 42		75 19		9 58 17	92 7	92	0 30	91 37	68 14	215	15	Slowed at Tunbridge Junction. Slowed on Cannon St. Bridge.
14	Cannon St.	Dover Town	11 7 a.m.	11 12 8	7 13 19	40 30	26 46		2 78		2 12 53	100 30	96	0 30	100 0	75 32	130	21½	Load after Ashford, 17 Coaches.
15	Cannon St.	Sandling	Sundays 11 7 a.m.	11 10 37	6 15 17	16 26	5 40		55 71		40 12 37	86 27	82	3 40	82 47	64 12	20½	16½	Slowed before Ashford, and stopped 1½ mins. there.
	UP.				Ashford.	Tunbridge Junction.	Halstead.	Chislehurst.	Tunbridge Junction.	New Cross.								
					M. S.	M. S.	M. S.	M. S.	M. S.	M. S.								
16	Shorncliffe	London Bridge	Sundays 3 4 p.m.	3 23	17 4	48 54	70 39	76 16	82 40		4 48 55	86 30	87	—	86 50	67 39	215	13	Signal check outside Cannon St. Heavy gale from west.
17	Sandling	Cannon St.	5 45 p.m.	6 35 56	13 16	47 13	68 59	74 45	81 49		8 3 20	87 44	84	1 30	86 14	64 18	172	18	Fog, north of Grove Park. Ran slowly, losing 1¼ mins.
18	Dover	Cannon St.	2 45 p.m.	3 1 41	29 48	59 45	81 15	87 15	95		4 43 41	102 0	102	1 30	100 30	75 32	60	16½	"Greasy" weather. * Distance is estimated. Folkestone Stn. is 69 m. 59 ch. Very slow running down bank from Halstead, losing over 4 minutes thus.
19	East of Folkestone	Cannon St.	4 10 p.m.	5 17 52	22	7 55	6 78	5 85	46 94		6 56 46	100 54	93	—	100 54	70*	24½	14	
20	Dover Town	anɪon St.	2 45 p.m.	3 4 32	29 43	61 48	84	4 90	0 97 12		4 47 57	103 15	102	0 15	103 0	75 32	24	14½	"Greasy" weather, 1½ min. lost by signal check near St. John's.

The runs which follow (79 in number) are not in all cases of importance. It was, however, thought better to include in the list many observations of little value as instances of fast express work, in order to illustrate the actual work of the line as regards semi-fast trains, and to disprove the frequently-expressed opinion that South-Eastern locomotives lose time in running. In order to facilitate reference all the runs are numbered and arranged in alphabetical order of departure stations. Those of importance have the initial number of the run in bolder type, and are marked with an asterisk.

OTHER FAST RUNS.

No. of Run.	From	To	Distance. M.	Distance. Ch.	Time due to Start.	Time Allowed. Min.	Time Taken. Min.	Time Taken. Sec.	Lost by Signal Checks. Min.	Lost by Signal Checks. Sec.	Net Time. Min.	Net Time. Sec.	Engine.	Load (Coaches).	Remarks.
21	Ashford	Paddock Wood	21	19	5 6 p.m.	29	27	32	—	—	27	32	19 inch	10	
22	,,	Sandling	9	27	7 3 p.m. (Sundays,	12	13	45	—	—	13	45	214	12½	
23	,,	,,	9	27	,,	12	14	34	—	—	14	34	19 inch	9	
24	,,	,,	9	27	,,	12	13	53	—	—	13	53	,,	7	
25	,,	,,	9	27	5 8 p.m.	17	15	56	—	—	15	56	199	16	
26	,,	Tunbridge Junc.			8 18 p.m.	40	35	57	—	—	35	57	19 inch	12½	
27	,,	,,			,,	40	37	3	2	30	34	33	203	8½	
*28	,,	,,			,,	40	34	26	—	—	34	26	241	16	
29	,,	,,			,,	40	38	21	—	—	38	21	199	11	
*30	,,	,,			,,	40	34	19	0	15	34	4	19 inch	15½	
31	,,	,,			,,	40	35	40	—	—	35	40	,,	12½	
32	,,	,,	26	42	,,	40	35	11	—	—	35	11	172	12	
*33	,,	,,			,,	40	33	22	—	—	33	22	201	12	
34	,,	,,			,,	40	38	0	—	—	38	0	203	16½	
35	,,	,,			,,	40	43	21	6	0	37	21	193	14	Note.—Attention is specially directed to Runs Nos. 28, 30, 33, 37, 39, 40, 43, 46, 48, 51, 62, 66 to 68, 74 to 77, 80 to 82, 84, 90, and 94.
36	,,	,,			8 53 p.m. (Sundays)	38	35	38	—	—	35	38	19 inch	15½	
*37	,,	,,			,,	38	35	31	—	—	35	31	172	18	
38	Cannon Street	Sandling			5 44 p.m.	87	88	15	3	30	84	45	78	7½	
*39	,,	,,			,,	87	85	54	2	0	83	54	19 inch	8	
*40	,,	,,			,,	87	88	18	5	0	83	18	214	8	
41	,,	,,			,,	87	88	37	3	0	85	37	19 inch	8½	
42	,,	,,	64	18	,,	87	89	15	4	45	84	30	,,	8	
*43	,,	,,			,,	87	86	36	3	0	83	36	,,	9	
44	,,	,,			,,	87	92	41	8	15	84	26	215	10	
45	,,	,,			,,	87	92	28	7	45	84	43	213	11	
*46	,,	Ashford	54	71	,, (Sundays)	76	76	57	5	30	71	27	214	12½	
47	,,	,,			,,	76	75	11	3	0	72	11	19 inch	6½	
*48	,,	,,			,,	76	76	52	6	0	70	52	,,	7	
49	,,	Tunbridge Junc.			5 44 p.m.	45	47	18	5	45	41	33	197	8½	
50	,,	,,			,,	45	45	11	3	15	41	56	197	9½	
*51	,,	,,	28	29	,,	45	46	29	6	0	40	29	116	13	
52	,,	,,			,,	45	45	0	1	0	44	0	197	12	
53	,,	,,			,,	45	51	19	10	15	41	4	116	8	
54	Dover Town	Folkestone			2 45 p.m. (Sundays)	10	10	3	—	—	10	3	197	12	
55	,,	,,	5	53	,,	10	10	57	—	—	10	57	215	13	
56	,,	,,			,,	10	10	20	—	—	10	20	140	13	
57	,,	Sandling	11	14	5 25 p.m.	19	19	52	—	—	19	52	172	18	
58	,,	,,	11	14	,,	19	19	37	—	—	19	37	139	15	
59	,,	Staplehurst	34	56	1 50 a.m.	49	47	0	—	—	47	0	19 inch	5	
60	Folkestone	Dover Town	5	53	9 39 a.m. (Sundays)	9	7	33	—	—	7	33	138	17½	

No. of Run.	From	To	Distance. M.	Distance. Ch	Time due to Start.	Time Allowed. Min.	Time Taken. Min.	Time Taken. Sec.	Lost by Signal Checks. Min.	Lost by Signal Checks. Sec.	Net Time. Min.	Net Time. Sec.	Engine.	Load (Coaches).	Remarks.
61	Folkestone	Dover Town	5	53	9 39 a.m. (Sundays)	9	8	6	—	—	8	6	241	18½	
*62	London Bridge	Ashford	54	16	3 38 p.m.	85	81	43	2	0	79	43	199	16	
63	"	Red Hill	20	55	5 25 p.m.	35	36	37	1	15	35	22	Old 6ft. cd.	12	
64	"	Tunbridge Junc.	27	54	4 45 p.m.	45	49	0	1	0	48	0	19 inch	21½	
65	New Cross.	Sevenoaks	17	11	—	—	34	0	0	45	33	15	"	22	This run is not included in Summary of Observations, the train being a "Relief," and not working to specified booked time.
*66	Sandling	Dover Town	11	14	12 30 p.m. (Sundays)	15	13	5	—	—	13	5	206	16½	Very smart short run.
*67	"	"	11	14	"	15	13	20	—	—	13	20	74	17	" "
*68	"	"	11	14	"	15	13	8	—	—	13	8	210	13½	" "
69	"	Radnor Park	4	50	7 12 p.m.	7	6	19	—	—	6	19	19 inch	8	
70	"	"	4	50	"	7	6	5	—	—	6	5	215	10	
71	Sevenoaks	Tunbridge Junc.	7	39	3 8 p.m.	12	14	17	3	0	11	17	19 inch	22	
72	"	"	7	39	"	12	10	51	—	—	10	51	241	17	
73	Staplehurst	"	12	27	2 41 a.m.	17	17	30	—	—	17	30	19 inch	5	
*74	Tunbridge Junc	Ashford	26	42	5 33 p.m.	35	33	39	—	—	33	39	199	13	
*75	"	"			"	35	36	20	—	—	36	20	204	19½	
*76	"	"			3 24 p.m.	37	37	20	—	—	37	20	19 inch	22	Severe gale.
*77	"	"			"	37	34	34	—	—	34	34	241	17	
78	"	Cannon Street	28	29	3 0 a.m.	45	42	0	—	—	42	0	19 inch	5	
79	"	Sandling	35	69	6 31 p.m.	45	45	34	—	—	45	34	197	8½	
*80	"	"			"	45	43	25	—	—	43	25	197	9½	
*81	"	"			"	45	45	39	—	—	45	39	116	13	
*82	"	"			"	45	44	4	—	—	44	4	197	12	
83	"	"			"	45	44	15	—	—	44	15	116	8	
*84	"	Sevenoaks	7	39	5 52 p.m.	15	13	45	—	—	13	45	19 inch	10	
85	"	"			9 3 p.m.	15	14	21	—	—	14	21	"	12½	
86	"	"			"	15	15	13	—	—	15	13	199	11	
87	"	"			"	15	16	8	—	—	16	8	19 inch	15½	
88	"	"			"	15	14	27	—	—	14	27	"	12½	
89	"	"			"	15	15	3	—	—	15	3	172	12	
*90	"	"			"	15	14	1	—	—	14	1	203	16½	
91	"	"			"	15	17	30	—	—	17	30	198	14	
92	"	"			9 20 p.m.	15	14	25	—	—	14	25	74	9	
93	"	"			"	15	16	8	—	—	16	8	19 inch	10½	
*94	"	"			"	15	14	27	—	—	14	27	241	16	
95	"	"			"	15	17	14	—	—	17	14	19 inch	22½	
96	"	"			"	15	16	19	—	—	16	19	241	14	
97	"	"			9 35 p.m.	15	15	54	—	—	15	54	19 inch	15½	
98	"	"			"	15	16	20	—	—	16	20	172	8	
99	"	"			"	15	15	18	—	—	15	18	19 inch	10	

LOCOMOTIVE WORK. 23

(b) CANTERBURY, RAMSGATE, AND MARGATE TRAINS.

	Run No. 100.	Run No. 101.		Run No. 102.	Run No. 103.
Train	1 38 p.m.	1 38 p.m.		7 50 p.m.	7 50 p.m.
From	New Cross	New Cross		Canterbury	Canterbury
To	Canterbury	Canterbury		New Cross	New Cross
Due to Arrive	3 0 p.m.	3 0 p.m.		9 19 p.m.	9 19 p.m.
Engine	130	130		130	130
Load (Coaches)	12¼	17		12¼	16¼
	H. M. S.	H. M. S.		H. M. S.	H. M. S.
Actual Departure	1 39 47	1 46 27		7 53 3	7 53 15

	Distance.	Time from Start.	Time from Start.		Distance.	Time from Start.	Time from Start.
	M. Ch.	Min. Sec.	Min. Sec.		M. Ch.	Min. Sec.	Min. Sec.
New Cross	—	—	—	Canterbury	—	—	—
St. John's	0 53	1 42	1 58	Chartham	3 15	6 6	5 55
Grove Park	4 3	6 55	8 36	Chilham	5 16	9 22	8 54
Chislehurst	6 31	10 53	13 1	Wye	9 77	15 59	15 11
Orpington	8 73	14 11	17 9	Ashford	14 20	21 48	21 15
Chelsfield	10 27	16 5	19 25	Pluckley	19 66	29 9	28 24
Halstead	11 50	17 59	21 52	Headcorn	25 8	34 49	34 3
Dunton Green	15 48	22 33	26 5¼	Staplehurst	28 35	38 27	37 29
Sevenoaks	17 11	24 21	28 31	Marden	30 76	41 30	40 28
Hildenborough	22 3	30 9	33 46	Paddock Wood	35 39	46 26	45 36
Tunbridge Junc.	24 50	32 59	36 20	Tunbridge Junc.	40 62	52 41	52 12
Paddock Wood	29 73	39 1	42 23	Hildenborough	43 29	56 27	56 6
Marden	34 36	43 34	46 58	Sevenoaks	48 21	63 57	64 52
Staplehurst	36 77	46 9	49 34	Dunton Green	49 64	65 42	66 44
Headcorn	40 24	49 25	52 55	Halstead	53 62	71 40	72 43
Pluckley	45 46	54 53	58 38	Chelsfield	55 5	73 14	74 13
Ashford	51 12	62 23	65 57	Orpington	56 39	74 49	75 44
Wye	55 35	70 38	72 23	Chislehurst	59 1	77 30	78 25
Chilham	60 16	83 36	77 48	Grove Park	61 29	79 56	80 48
Chartham	62 17	85 57	81 3	St. John's	64 59	83 50	84 28
Canterbury	65 32	89 48	86 1	New Cross	65 32	85 6	85 28

		H. M. S.	H. M. S.		H. M. S.	H. M. S.
Actual Arrival		3 9 35	3 12 28		9 17 9	9 18 43
Time Taken		89 48	86 1		85 6	85 28
Signal Delays		12 0	4 45		—	0 15
Net Time		77 48	81 16		85 6	85 13
Time Allowed		82 0	82 0		89 0	89 0
Gain or Loss		Gain 4 mins. 12 secs.	Gain 44 seconds		Gain 3 mins. 54 secs.	Gain 3 mins. 47 secs.
Particulars of Signal Delays, etc.		Three severe Signal Checks between Pluckley and Chilham and stop of 3¼ mins. near Wye.	Three severe Signal Checks between Pluckley and Canterbury.			Slight Check, losing ¼ min. just before Ashford.

We have, in the four examples detailed above, probably our finest observations of South Eastern work. The uphill running of No. 100 from St. John's to Halstead and of No. 102 from Tunbridge Junction to Sevenoaks is especially good, and the work of the former up the bank will be afterwards given in detail, while the running on the level of both the down trains (and especially of No. 101) from Paddock Wood to Pluckley is as good as is met with on any line in England. With a load of 17 Coaches it is a performance of great merit to cover 15¾ miles in 16¼ minutes with grades on the whole slightly adverse.

THE SOUTH EASTERN RAILWAY.

No. of Run.	From	To	Distance. M. Ch.	Time due to Start.	Time Allowed. Mins.	Time Taken. Min. Sec.	Lost by Signal Check. Min. Sec.	Net Time. Min. Sec.	Engine.	Load. Coaches	Remarks.
104	Ashford	Canterbury	14 20	6 28 p.m.	20	19 1	—	19 1	204	12½	
105	,,	,,	14 20	11 30 a.m. (Sundays)	20	18 11	0 15	17 56	205	7½	
*106	,,	New Cross	51 12	5 56 p.m. (Sundays)	67	66 3	0 15	65 48	205	7½	
107	Canterbury	Ashford	14 20	5 32 p.m. (Sundays)	22	21 7	0 45	20 22	205	7½	
108	,,	Minster		3 2 p.m.	17	15 27	—	15 27	130	12½	} Very smart short run.
*109	,,	,,	11 39	,,	17	14 16	—	14 16	130	16½	
110	,,	,,		11 50 a.m. (Sundays)	17	14 26	—	14 26	205	7½	
*111	,,	New Cross	65 32	7 50 p.m.	89	97 36	11 30	86 6	94	15½	
112	Minster	Canterbury	11 39	7 30 p.m.	18	16 15	—	16 15	130	16½	
113	,,	,,	11 39	5 13 p.m. (Sundays)	19	16 3	—	16 3	205	7½	
*114	New Cross	Ashford	51 12	10 20 a.m. (Sundays)	65	75 2	10 0	65 2	205	7½	NOTE.—Attention is specially directed to Runs Nos. 106, 109, 111, 114, and 115.
*115	,,	Canterbury	65 32	1 38 p.m.	82	88 23	3 45	84 38	43	12	
116	Ramsgate	Minster	4 30	7 20 p.m.	8	6 36	—	6 36	130	14	
117	Sevenoaks	New Cross	17 11	9 51 p.m.	31	27 0	4 30	22 30	19 inch	15½	
118	,,	,,	17 11	,, (Sundays)	31	23 47	—	23 47	172	18	

Plate II.

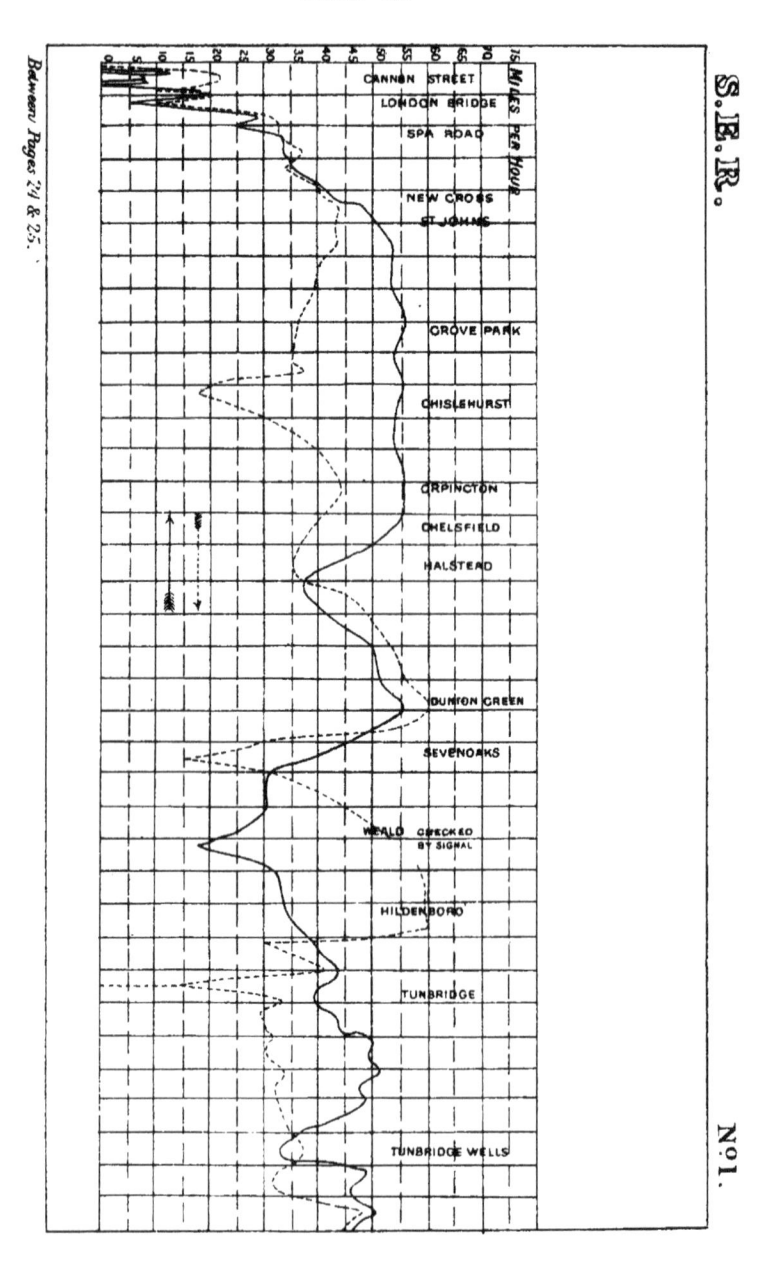

SPEED RECORDER DIAGRAM.

S.E.R.

No. 1.

Plate III.

SPEED RECORDER DIAGRAM.

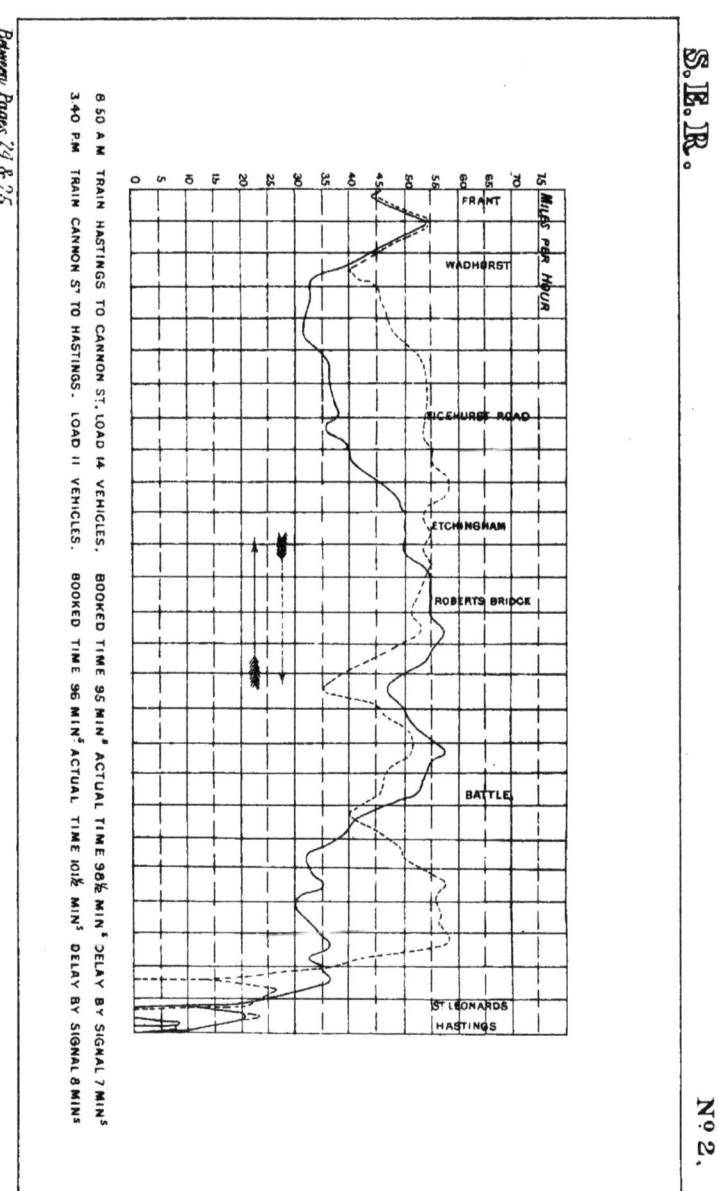

S.E.R.

No 2.

(c) HASTINGS TRAINS.

Runs Nos. 121 and 122 are given in the form of Speed Recorder Diagrams.

	Run No. 119.		Run No. 120.
Train From To Due to arrive Engine Load (Coaches) Actual Departure.	3 50 p.m. Cannon St. West St. Leonards 5 18 p.m. 91 12 H. M. S. 3 52 8		8 48 a.m. St. Leonards Cannon St. 10 25 a.m. 79 15 H. M. S. 8 51 45

	Distance.	Time from Start.		Distance.	Time from Start.
	M. Ch.	Min. Sec.		M. Ch.	Min. Sec.
Cannon Street London Bridge Spa Road New Cross St. John's Grove Park Chislehurst Orpington Chelsfield Halstead Dunton Green Sevenoaks Hildenborough Tunbridge Junc. Southborough Tunbridge Wells Frant Wadhurst Ticehurst Road Etchingham Robertsbridge Battle West St. Leonards	 0 55 1 50 3 59 4 32 7 62 10 10 12 52 14 6 15 29 19 27 20 70 25 62 28 29 31 58 33 17 35 41 38 12 42 55 46 22 48 37 54 31 59 46	 1 29 3 3 5 56 6 59 14 22 19 0 22 48 24 46 26 56 33 14 35 24 42 2 44 51 51 26 54 23 57 54 60 59 65 55 69 45 72 14 80 27 87 22	St. Leonards West St. Leonards. Battle Robertsbridge Etchingham Ticehurst Road Wadhurst Frant Tunbridge Wells Southborough Tunbridge Junc. Hildenborough Sevenoaks Dunton Green Halstead Chelsfield Orpington Chislehurst Grove Park St. John's New Cross Spa Road London Bridge Cannon Street	 0 79 6 14 12 8 14 23 17 70 22 33 25 4 27 28 28 67 32 16 34 63 39 55 41 18 45 16 46 39 47 73 50 35 52 63 56 13 56 66 58 75 59 70 60 45	 2 12 3 19 0 21 31 26 46 35 22 38 52 42 0 44 24 48 47 53 40 65 37 67 32 73 20 74 58 76 27 79 6 81 55 86 10 87 4 89 55 92 23 97 5
Actual Arrival Time Taken Signal Delays Net Time Time Allowed Gain or Loss Particulars of Signal Delays, etc.		H. M. S. 5 19 30 87 22 4 15 83 7 88 0 Gain 4 mins. 53 secs. Signals north of Grove Park, 1¼ mins.; also at Halstead and Sevenoaks 2¾ mins.			H. M. S. 10 28 50 97 5 7 45 89 20 97 0 Gain 7 mins. 40 secs. Signals at Hildenboro', 3¾ mins.; at Spa Road, 1 min.; at London Bridge (including ½ min. stop), 3 mins.

Although a route of very stiff grades throughout, it would appear, from the above observations and from Run 126, briefly referred to in next table, that, could signal checks be avoided, the best Hastings expresses are not very hardly timed. The speed up the banks is, however, distinctly good and equal to the average of that attained by the Continental trains. We are able, through the kindness of Sir Myles Fenton, to give speed recorder diagrams of two excellent performances bearing out these remarks.

THE SOUTH EASTERN RAILWAY.

No. of Run.	From	To	Distance.	Time due to Start.	Time Allowed	Time Taken	Lost by Signal Checks.	Net Time	Engine.	Load.	Remarks
			M. Ch.		Min.	Min. Sec.	Min. Sec.	Min. Sec.		Coaches	
123	Cannon Street	Tunbridge Wells	33 17	5 0 p.m.	54	58 2	2 45	55 17	19 inch	13	
124	,,	,,	33 17	,,	54	57 0	4 15	52 45	,,	13	
125	,,	,,	33 17	,,	54	58 22	3 15	55 7	149	13	
*126	,,	West St. Leonards	59 46	3 50 p.m.	88	95 5	12 45	82 20	79	12	Fine performance.
127	Chislehurst	Sevenoaks	11 15	6 10 a.m.	20	18 30	—	18 30	19 inch	7	
128	London Bridge	Chislehurst	9 0	5 53 a.m.	16	17 30	1 30	16 0	,,	7	
129	Sevenoaks	Tunbridge Junc.	7 39	6 31 a.m.	12	11 45	—	11 45	,,	7	
130	Tunbridge Junc.	Sevenoaks	7 39	—	15	15 18	—	15 18	19 inch	9½	
131	Tunbridge Wells	New Cross	29 38	8 8 p.m.	48	46 2	—	46 2	205	11	
132	,,	,,	29 38	,,	48	44 11	—	44 11	60	10½	
133	,,	Wadhurst	4 75	5 57 p.m.	10	9 4	—	9 4	19 inch	13	
134	,,	,,	4 75	,,	10	8 52	—	8 52	,,	13	
135	,,	,,	4 75	,,	10	9 30	—	9 30	149	13	
136	,,	West St. Leonards	26 29	7 5 a.m.	40	38 45	—	38 45	—	7	
137	Wadhurst	,,	21 34	6 8 p.m.	32	{23 14 / 8 0}	—	31 16	19 inch	13	{Train stopped at Battle.
138	,,	,,	21 34	,,	32	{22 0 / 8 38}	—	30 38	,,	13	,,
139	,,	,,	21 34	,,	32	{23 40 / 8 5}	—	31 45	149	13	,,
140	West St. Leonards	Tunbridge Wells	26 29	7 23 p.m.	42	44 11	—	44 11	205	11	
141	,,	,,	26 29	,,	42	42 15	—	42 15	60	10½	

(d) CHATHAM TRAINS.

	Run No. 142.		No. 143.	No. 144.
Train	5 41 p.m.		10 18 p.m.	10 18 p.m.
From	London Bridge.		Gravesend	Gravesend
To	Dartford.		London Bridge.	London Bridge.
Due to Arrive	6 5 p.m.		10 51 p.m.	10 51 p.m.
Engine	6 ft. coupled Bogie.		88 (Old 6 ft. Coupled.)	58 (Tank)
Load (Coaches)	13		9	9½
	H. M. S.		H. M. S.	H. M. S.
Actual Departure	5 46 38		10 18 4	10 19 0

	Distance.	Time from Start.		Distance.	Time from Start.	Time from Start.
	M. Ch.	Min. Sec.		M. Ch.	Min. Sec.	Min. Sec.
London Bridge			Gravesend			
Spa Road	0 75	2 3	Northfleet	2 10	4 2	4 10
New Cross	3 4	6 35	Greenhithe	4 11	6 51	7 3
St. John's	3 57	7 48	Dartford	6 65	10 13	11 47
Lee	5 72	11 35	Crayford	8 51	12 37	13 28
Eltham	7 42	14 32	Bexley	10 6	14 34	15 29
New Eltham	8 38	15 52	Sidcup	12 1	17 39	18 31
Sidcup	10 2	17 50	New Eltham	13 45	19 58	20 34
Bexley	11 77	20 4	Eltham	14 41	21 14	21 44
Crayford	13 32	21 48	Lee	16 11	23 33	23 47
Dartford	15 18	24 9	St. John's	18 26	26 20	26 47
			New Cross	18 79	27 13	27 43
			Spa Road	21 8	29 57	30 29
			London Bridge	22 3	31 47	32 12

	H. M. S.		H. M. S.	H. M. S.
Actual Arrival	6 10 47		10 49 51	10 51 12
Time Taken	24 9		31 47	32 12
Signal Delays	1 45		0 15	—
Net Time	22 24		31 32	32 12
Time Allowed	24 0		33 0	33 0
Gain or Loss	Gain 1 min. 36 secs.		Gain 1 min. 28 secs.	Gain 48 secs.
Particulars of Signal Delays, etc.	1¾ mins. lost by Signal Check north of New Cross.		¼ min. lost by Check outside London Bridge.	

THE SOUTH EASTERN RAILWAY.

No. of Run.	From	To	Distance. M. Ch.	Time due to Start.	Time Allowed. Min.	Time Taken. Min. Sec.	Lost by Signal Checks. Min. Sec.	Net Time. Min. Sec.	Engine.	Load. Coaches	Remarks.
*145	Gravesend	London Bridge	22 3	10 18 p.m.	33	38 0	6 45	31 15	146 (Tank)	9	NOTE.—Attention is specially directed to Runs Nos. 145 and 149 to 151.
146	,,	Strood	} 7 20	6 20 p.m.	12	10 47	—	10 47	Old 6ft. Cpld.	13	
147	,,	,,		,,	12	10 20	—	10 20	6ft. Cpld. Bogie	13	
148	,,	,,		8 2 p.m.	12	11 26	—	11 26	58 (Tank)	10	
*149	London Bridge	Dartford	} 15 18	5 41 p.m.	24	23 36	0 30	23 6	Old 6ft. Cpld.	13	
*150	,,	,,		,,	24	26 15	3 15	23 0	6ft. Cpld. Bogie	13	
*151	,,	,,		,,	24	25 11	2 30	22 41	6ft. Cpld. Bogie	13	
152	,,	,,		,,	24	28 42	4 15	24 27	6ft. Cpld. Bogie	13	
153	,,	,,		,,	24	30 8	6 15	23 53	Old 6ft. Cpld.	13	
154	Strood	Gravesend	7 20	10 5 p.m.	12	12 32	—	12 32	58 (Tank)	10	

Having now, perhaps, more than sufficiently illustrated South Eastern working of fast trains in actual practice, it only remains to offer a few remarks on the subjects of (1) downhill and (2) uphill running, and, in conclusion, to present a table giving our best performances between each pair of stations on the main line, and to summarise our experiences.

(1) *DOWNHILL RUNNING.*—Great caution is exercised in running down grades on this railway, sixty miles an hour being rarely exceeded, and, even on some occasions, scarcely attained. There is little oscillation, and the rolling stock travels very smoothly, the only exceptions being, perhaps, while rounding the curves of the Hastings line and those on approaching Sevenoaks from the north. On referring to the subjoined table showing the best times recorded between each pair of stations on the main line, it will be seen that if the practice of running at seventy-five to eighty miles an hour in vogue on down grades on the Midland, Great Northern, and other lines were adopted by the South Eastern, much better time could be made than at present.

(2) *UPHILL RUNNING.*—Although in the tables given above showing actual performances the times of passing the stations only are stated, yet in nearly all cases milepost timings up the steep grades between St. John's and Halstead, Tunbridge Junction and Sevenoaks, and on the Hastings line were made. Some of these are given below, and, in most cases, show good work. Probably the best instance is that of engine 130 and 12½ coaches; but engine 156 with 20 on, engine 138 with 19 on, and engine 241 with 18½ on, also afforded remarkable performances, while the heavy load of 21½ coaches is clear proof that on these severe grades the locomotive is by no means spared. No instances are given below except for the bank from St. John's to Halstead, as on the other similar banks on the line the speeds attained with like loads were almost identical. In a general way the best of these results equal—with the exception, perhaps, of some of the crack Scotch trains—the work done on up grades by the express lines north of the Thames, and are, therefore, to be accounted satisfactory. If comparison be desired, readers should turn to the chapter on *Uphill Running* in the Appendix to "British Railways," where some of the results obtained by numerous and careful observations on the writer's part are set forth.

UPHILL RUNNING. NEW CROSS TO HALSTEAD.

Between Mileposts			4 and 5	5 and 6	6 and 7	7 and 8	8 and 9	9 and 10	10 and 11	11 and 12	12 and 13	13 and 13¾	13¾ & 14½	14½ & 15¼	
Rate of Gradient			1 in 180 and 1 in 250	Chiefly 1 in 140	1 in 140 and 1 in 120	1 in 120	1 in 120 and short stretch of level	Level and 1 in 146	1 in 231	1 in 234 level, and 1 in 310	1 in 310 and 1 in 120	1 in 120	1 in 120	1 in 120 and 1 in 170	
Up or Down			Up	Up	Up	Up	Up	Up	Up	Up	Up	Up	Up	Up	
Engine.	Load.	Reference to No. of Run from which this is extracted.	Speed in Miles per Hour.												Remarks.
91	12	119	Checked by Signals				30	38	40	45	44	41	36	—	
205	7½	114	34	36	36	37	36	Signal	Check	41	42	Signal Check			Started from New Cross.
130	21½	14	41	35	29	24	23	29	31	34	34	28	23	23	"Greasy" Weather.
74	17	2	38	35	32	31	30	35	36	40	40	34	30	30	
156	20	3	40	36	31	27	26	31	33	38	37	33	28	29	
210	13½	1	40	37	33	31	31	37	37	40	40	36	32	31	
43	12	115	—	36	34	33	33	37	39	42	41	37	33	32	Started from New Cross.
79	12	126	Signal Checks			30	29	Signal	Check	40	42	36	33	33	
206	16½	15	40	36	33	31	31	34	34	39	40	36	33	33	
214	12½	46	41	38	34	31	29	34	36	40	41	36	32	32	
130	12½	100	38	41	38	39	37	42	41	44	46	42	41	40	Started from New Cross.
138	19	6	—	38	34	31	29	33	34	39	39	34	30	30	
187	16	—	—	—	—	33	31	37	37	41	39	34	30	30	
241	18½	5	43	37	32	29	27	34	36	38	40	35	32	31	
130	17	101	31	31	30	31	31	35	36	40	40	33	31	31	Started from New Cross.
2	15	4	42	37	32	29	27	33	34	38	38	34	31	31	

NOTE.—The 4th Milepost is nearly ¼-mile south of St. John's Station.

BEST PERFORMANCES.—The best times noted between each pair of stations on the South Eastern main line will be found summarised below.

BEST PERFORMANCES

BETWEEN PAIRS OF STATIONS ON MAIN LINE.

Distance.	Between	And	Best Time.	Load.	Between	And	Best Time.	Load.	Distance.
M. Ch.			Min. Sec.	Coaches.			Min. Sec.	Coaches.	M. Ch.
0 55	Cannon Street	London Bridge	1 21	6½	Dover Town	Folkestone	10 3	12	5 53
0 75	London Bridge	Spa Road	1 34	12	Folkestone	Radnor Park	1 29	14½	0 71
2 9	Spa Road	New Cross	2 49	19	Radnor Park	Shorncliffe	1 1	14½	0 54
0 53	New Cross	St. John's	0 54	18½	Shorncliffe	Sandling	5 31	14½	3 76
3 30	St. John's	Grove Park	5 13	19	Sandling	Westenhanger	1 40	16½	1 21
2 28	Grove Park	Chislehurst	3 38	12½	Westenhanger	Smeeth	4 10	13	3 61
2 42	Chislehurst	Orpington	3 33	12½	Smeeth	Ashford	4 31	13	4 25
1 34	Orpington	Chelsfield	1 54	12½	Ashford	Pluckley	6 17	16½	5 46
1 23	Chelsfield	Halstead	1 54	12½	Pluckley	Headcorn	5 33	16½	5 22
3 78	Halstead	Dunton Green	4 34	12½	Headcorn	Staplehurst	3 26	16½	3 27
1 43	Dunton Green	Sevenoaks	1 36	13	Staplehurst	Marden	2 50	13	2 41
4 72	Sevenoaks	Hildenborough	5 15	17	Marden	Paddock Wood	4 56	12½	4 43
2 47	Hildenborough	Tunbridge Junc.	2 28	13½	Paddock Wood	Tunbridge Junc.	6 15	12½	5 23
5 23	Tunbridge Junc.	Paddock Wood	5 56	17	Tunbridge Junc	Hildenborough	3 37	13	2 47
4 43	Paddock Wood	Marden	4 33	12½	Hildenborough	Sevenoaks	7 30	12½	4 72
2 41	Marden	Staplehurst	2 35	12½	Sevenoaks	Dunton Green	1 45	13	1 43
3 27	Staplehurst	Headcorn	3 16	12½	Dunton Green	Halstead	5 2	13	3 78
5 22	Headcorn	Pluckley	5 28	12½	Halstead	Chelsfield	1 20	13	1 23
5 46	Pluckley	Ashford	6 16	13½	Chelsfield	Orpington	1 25	13	1 34
4 25	Ashford	Smeeth	4 44	13½	Orpington	Chislehurst	2 26	16½	2 42
3 61	Smeeth	Westenhanger	4 30	15½	Chislehurst	Grove Park	2 14	16½	2 28
1 21	Westenhanger	Sandling	1 33	15½	Grove Park	St. John's	3 20	13	3 30
3 76	Sandling	Shorncliffe	3 54	15½	St. John's	New Cross	0 49	16½	0 53
0 54	Shorncliffe	Radnor Park	0 42	13½	New Cross	Spa Road	2 41	13½	2 9
0 71	Radnor Park	Folkestone	0 53	16½	Spa Road	London Bridge	1 29	14½	0 75
5 53	Folkestone	Dover Town	5 56	17	London Bridge	Cannon Street	1 37	14	0 55
			83 4				90 17		

SUMMARY OF OBSERVATIONS

Number of runs made ...	
Gross time allowed ...	
Actual time taken including Signal Checks and other Detentions	
Loss on Booked Time	
Percentage of Total Time allowed	**Loss**
Time Lost by Signal Checks and other Detentions	
Net Time (allowing for Signal Checks and other Detentions)	
Gain on Booked Time	
Percentage of Total Time allowed	**Gain**